Maintenance Log Book

Date / Time	Maintenance / Comments	

Maintenance Log Book

Log Book # __________

Date / Time	Maintenance / Comments	Signature

Maintenance Log Book

Log Book # __________

Date / Time	Maintenance / Comments	Signature

Maintenance Log Book

Log Book # __________

Date / Time	Maintenance / Comments	Signature

Maintenance Log Book

Log Book # __________

Date / Time	Maintenance / Comments	Signature

Maintenance Log Book

Log Book # __________

Date / Time	Maintenance / Comments	Signature

Maintenance Log Book

Log Book # __________

Date / Time	Maintenance / Comments	Signature

Maintenance Log Book

Log Book # __________

Date / Time	Maintenance / Comments	Signature

Maintenance Log Book

Log Book # __________

Date / Time	Maintenance / Comments	Signature

Maintenance Log Book

Log Book # __________

Date / Time	Maintenance / Comments	Signature

Maintenance Log Book

Log Book # __________

Date / Time	Maintenance / Comments	Signature

Maintenance Log Book

Log Book # __________

Date / Time	Maintenance / Comments	Signature

Maintenance Log Book

Log Book # __________

Date / Time	Maintenance / Comments	Signature

Maintenance Log Book

Log Book # __________

Date / Time	Maintenance / Comments	Signature

Maintenance Log Book

Log Book # __________

Date / Time	Maintenance / Comments	Signature

Maintenance Log Book

Log Book # __________

Date / Time	Maintenance / Comments	Signature

Maintenance Log Book

Log Book # __________

Date / Time	Maintenance / Comments	Signature

Maintenance Log Book

Log Book # __________

Date / Time	Maintenance / Comments	Signature

Maintenance Log Book

Log Book # __________

Date / Time	Maintenance / Comments	Signature

Maintenance Log Book

Log Book # __________

Date / Time	Maintenance / Comments	Signature

Maintenance Log Book

Log Book # __________

Date / Time	Maintenance / Comments	Signature

Maintenance Log Book

Log Book # __________

Date / Time	Maintenance / Comments	Signature

Maintenance Log Book

Log Book # __________

Date / Time	Maintenance / Comments	Signature

Maintenance Log Book

Log Book # __________

Date / Time	Maintenance / Comments	Signature

Maintenance Log Book

Log Book # __________

Date / Time	Maintenance / Comments	Signature

Maintenance Log Book

Log Book # __________

Date / Time	Maintenance / Comments	Signature

Maintenance Log Book

Log Book # __________

Date / Time	Maintenance / Comments	Signature

Maintenance Log Book

Log Book # __________

Date / Time	Maintenance / Comments	Signature

Maintenance Log Book

Log Book # __________

Date / Time	Maintenance / Comments	Signature

Maintenance Log Book

Log Book # __________

Date / Time	Maintenance / Comments	Signature

Maintenance Log Book

Log Book # __________

Date / Time	Maintenance / Comments	Signature

Maintenance Log Book

Log Book # __________

Date / Time	Maintenance / Comments	Signature

Maintenance Log Book

Log Book # __________

Date / Time	Maintenance / Comments	Signature

Maintenance Log Book

Log Book # ________

Date / Time	Maintenance / Comments	Signature

Maintenance Log Book

Log Book # __________

Date / Time	Maintenance / Comments	Signature

Maintenance Log Book

Log Book # __________

Date / Time	Maintenance / Comments	Signature

Maintenance Log Book

Log Book # __________

Date / Time	Maintenance / Comments	Signature

Maintenance Log Book

Log Book # __________

Date / Time	Maintenance / Comments	Signature

Maintenance Log Book

Log Book # __________

Date / Time	Maintenance / Comments	Signature

Maintenance Log Book

Log Book # __________

Date / Time	Maintenance / Comments	Signature

Maintenance Log Book

Log Book # __________

Date / Time	Maintenance / Comments	Signature

Maintenance Log Book

Log Book # __________

Date / Time	Maintenance / Comments	Signature

Maintenance Log Book

Log Book # __________

Date / Time	Maintenance / Comments	Signature

Maintenance Log Book

Log Book # __________

Date / Time	Maintenance / Comments	Signature

Maintenance Log Book

Log Book # __________

Date / Time	Maintenance / Comments	Signature

Maintenance Log Book

Log Book # __________

Date / Time	Maintenance / Comments	Signature

Maintenance Log Book

Log Book # __________

Date / Time	Maintenance / Comments	Signature

Maintenance Log Book

Log Book # __________

Date / Time	Maintenance / Comments	Signature

Maintenance Log Book

Log Book # __________

Date / Time	Maintenance / Comments	Signature

Maintenance Log Book

Log Book # __________

Date / Time	Maintenance / Comments	Signature

Maintenance Log Book

Log Book # __________

Date / Time	Maintenance / Comments	Signature

Maintenance Log Book

Log Book # __________

Date / Time	Maintenance / Comments	Signature

Maintenance Log Book

Log Book # __________

Date / Time	Maintenance / Comments	Signature

Maintenance Log Book

Log Book # __________

Date / Time	Maintenance / Comments	Signature

Maintenance Log Book

Log Book # __________

Date / Time	Maintenance / Comments	Signature

Maintenance Log Book

Log Book # __________

Date / Time	Maintenance / Comments	Signature

Maintenance Log Book

Log Book # __________

Date / Time	Maintenance / Comments	Signature

Maintenance Log Book

Log Book # __________

Date / Time	Maintenance / Comments	Signature

Maintenance Log Book

Log Book # __________

Date / Time	Maintenance / Comments	Signature

Maintenance Log Book

Log Book # __________

Date / Time	Maintenance / Comments	Signature

Maintenance Log Book

Log Book # __________

Date / Time	Maintenance / Comments	Signature

Maintenance Log Book

Log Book # __________

Date / Time	Maintenance / Comments	Signature

Maintenance Log Book

Log Book # __________

Date / Time	Maintenance / Comments	Signature

Maintenance Log Book

Log Book # __________

Date / Time	Maintenance / Comments	Signature

Maintenance Log Book

Log Book # __________

Date / Time	Maintenance / Comments	Signature

Maintenance Log Book

Log Book # __________

Date / Time	Maintenance / Comments	Signature

Maintenance Log Book

Log Book # __________

Date / Time	Maintenance / Comments	Signature

Maintenance Log Book

Log Book # __________

Date / Time	Maintenance / Comments	Signature

Maintenance Log Book

Log Book # __________

Date / Time	Maintenance / Comments	Signature

Maintenance Log Book

Log Book # __________

Date / Time	Maintenance / Comments	Signature

Maintenance Log Book

Log Book # __________

Date / Time	Maintenance / Comments	Signature

Maintenance Log Book

Log Book # __________

Date / Time	Maintenance / Comments	Signature

Maintenance Log Book

Log Book # __________

Date / Time	Maintenance / Comments	Signature

Maintenance Log Book

Log Book # __________

Date / Time	Maintenance / Comments	Signature

Maintenance Log Book

Log Book # __________

Date / Time	Maintenance / Comments	Signature

Maintenance Log Book

Log Book # __________

Date / Time	Maintenance / Comments	Signature

Maintenance Log Book

Log Book # __________

Date / Time	Maintenance / Comments	Signature

Maintenance Log Book

Log Book # __________

Date / Time	Maintenance / Comments	Signature

Maintenance Log Book

Log Book # __________

Date / Time	Maintenance / Comments	Signature

Maintenance Log Book

Log Book # __________

Date / Time	Maintenance / Comments	Signature

Maintenance Log Book

Log Book # __________

Date / Time	Maintenance / Comments	Signature

Maintenance Log Book

Log Book # __________

Date / Time	Maintenance / Comments	Signature

Maintenance Log Book

Log Book # __________

Date / Time	Maintenance / Comments	Signature

Maintenance Log Book

Log Book # __________

Date / Time	Maintenance / Comments	Signature

Maintenance Log Book

Log Book # __________

Date / Time	Maintenance / Comments	Signature

Maintenance Log Book

Log Book # __________

Date / Time	Maintenance / Comments	Signature

Maintenance Log Book

Log Book # __________

Date / Time	Maintenance / Comments	Signature

Maintenance Log Book

Log Book # __________

Date / Time	Maintenance / Comments	Signature

Maintenance Log Book

Log Book # __________

Date / Time	Maintenance / Comments	Signature

Maintenance Log Book

Log Book # __________

Date / Time	Maintenance / Comments	Signature

Maintenance Log Book

Log Book # __________

Date / Time	Maintenance / Comments	Signature

Maintenance Log Book

Log Book # __________

Date / Time	Maintenance / Comments	Signature

Maintenance Log Book

Log Book # __________

Date / Time	Maintenance / Comments	Signature

Maintenance Log Book

Log Book # __________

Date / Time	Maintenance / Comments	Signature

Maintenance Log Book

Log Book # __________

Date / Time	Maintenance / Comments	Signature

Maintenance Log Book

Log Book # __________

Date / Time	Maintenance / Comments	Signature

Maintenance Log Book

Log Book # __________

Date / Time	Maintenance / Comments	Signature

Maintenance Log Book

Log Book # __________

Date / Time	Maintenance / Comments	Signature

Maintenance Log Book

Log Book # __________

Date / Time	Maintenance / Comments	Signature

Maintenance Log Book

Log Book # __________

Date / Time	Maintenance / Comments	Signature

Maintenance Log Book

Log Book # __________

Date / Time	Maintenance / Comments	Signature

Maintenance Log Book

Log Book # __________

Date / Time	Maintenance / Comments	Signature

Maintenance Log Book

Log Book # __________

Date / Time	Maintenance / Comments	Signature

Maintenance Log Book

Log Book # __________

Date / Time	Maintenance / Comments	Signature

Maintenance Log Book

Log Book # __________

Date / Time	Maintenance / Comments	Signature

Maintenance Log Book

Log Book # __________

Date / Time	Maintenance / Comments	Signature

Maintenance Log Book

Log Book # __________

Date / Time	Maintenance / Comments	Signature

Maintenance Log Book

Log Book # __________

Date / Time	Maintenance / Comments	Signature

Maintenance Log Book

Log Book # ________

Date / Time	Maintenance / Comments	Signature

Maintenance Log Book

Log Book # __________

Date / Time	Maintenance / Comments	Signature

Made in United States
Troutdale, OR
03/02/2024

18150304R00062